THE LITTLE
GREEN
HANDBOOK

145 Simple Steps to Save the Planet

RUTH CULLEN

ILLUSTRATED BY
KERREN BARBAS STECKLER

PETER PAUPER PRESS, INC.
WHITE PLAINS, NEW YORK

For my dear husband, Jay
Special thanks to Kerry Laberge for leading "green"
by example, and to the bright green students in
Mrs. Gauthier's third grade class at Union School

Designed by Heather Zschock
Illustrations © 2008 Kerren Barbas Steckler

Copyright © 2008
Peter Pauper Press, Inc.
202 Mamaroneck Avenue
White Plains, NY 10601
All rights reserved
ISBN 978-1-59359-829-7
Printed in Hong Kong
7 6 5 4 3

Visit us at www.peterpauper.com

THE LITTLE

GREEN
HANDBOOK

Contents

introduction

Let's be practical. You're busy. I'm busy. Saving the planet has to fit in between work and wine tasting. It has to be simple and easy, especially if it's going to stick. It's not that we're la-a-a-azy about these things, or addicted to a high-maintenance lifestyle . . . well, maybe just a little. But let's work with that.

Here is your treasure trove of simple actions you can take and green choices you can make without turning your life upside down.

We worry, and rightfully so, about the world. Talk about global warming can send a chill right down your spine. And then there are the overflowing landfills, pollution, and our natural resources drying up.

It's easy to feel overwhelmed by the scope and magnitude of the environmental challenges facing our planet. It's easy to wonder how—*or if*—our actions can make a difference. It's easy to feel discouraged and powerless.

But it's also easy—contrary to popular opinion—being green. *The Little Green Handbook* can help you turn your fears into actions, one simple green step at a time. Whether you're at home, in the workplace, or out and about, you can make smart, earth-friendly choices that respect the planet and sustain life.

Green may be the new black, but it's so much more than a color or trend. It's a mind-set. A lifestyle. A movement. But it's also down-to-earth. It's the stuff of this little book, like knowing where and how to recycle old sneakers or cell phones, which fruits and veggies are the most pesticide-prone, which plastics are bad for you, how to garden naturally, where to find biodegradable trash bags, how to save energy by taking simple actions like dusting your lightbulbs, and which houseplants freshen air most effectively.

And then there are bigger-picture items, part of the whole conservation conversation, that involve educating yourself and your world, using consumer power to effect change, and speaking out.

So take heart, take action, and have a green toast to our little green planet . . . with *organic* wine, of course!

home green home

*We do not inherit the Earth
from our ancestors, we borrow
it from our children.*

—NATIVE AMERICAN PROVERB

There's never been a better time than now for an extreme *green* home makeover!

Your small changes on the home front— from the way you sort your garbage to the way you manage your consumption of water, electricity, and fuel—can add up to big savings for you and the environment.

Rethink your approach to home maintenance. Examine the products you use to clean your home and fertilize your lawn. Notice when the water is running, and when it's not.

Do a quick inventory of your lighting, appliances, and electronics, and pay close attention to things like on/off switches and electrical outlets.

A home energy audit will help you identify even more ways to make your home a lean, *green* running machine.

It's easier than you think. Start small and do one green thing at home. You might be surprised at how quickly your small changes make a big difference.

fill your recycling bin

Each week, the average person produces a *staggering* amount of trash—much of it composed of recyclable materials that end up at the town dump.

If you're among the nearly 25 percent of people who don't recycle, consider this: recycling just *one* aluminum can saves enough energy to power a television for *three* hours.

Imagine what your cans, plastic bottles, cardboard boxes, glass jars, and newspapers could do!

Be part of the solution, and don't let laziness prevent you from doing the right thing.

Take baby steps toward the recycling bin provided by your city or town, and *start recycling now.*

know your recyclables

Before you toss that glass jar or children's toy in the trash, ask yourself one question: *can this be recycled?* You might be surprised by the answer.

Consult your local sanitation or public works department for specific recycling guidelines in your area, or visit *http://earth911.org* and plug in your zip code.

know your recyclables

Here is a partial list of common
recyclables to help get you started.
(Check your local laws.
They may differ.)

- paper: newspapers; phone books;
  magazines; junk mail; computer paper

- cardboard: cereal and shoe boxes;
  cardboard packaging; toilet paper/
  paper towel tubes

- aluminum: cans; clean aluminum foil

- glass: bottles; jars; condiment and spice
  containers

- plastic: soda and water bottles; milk
  jugs; in general, any containers marked
  ♲ or ♴ on the bottom

unscrew your caps

 Remove the screw tops from your glass and plastic bottles before tossing them into the recycling bin.

Most bottles feature a combination of materials that must be recycled separately to avoid cross-contamination. By removing the caps from the bottles, you can help streamline and improve the entire recycling process.

And remember: not all plastics are created equally.

The plastic cap on your water bottle, for example, is made from a different plastic than the water bottle itself. One or the other may not be recyclable in your town. Always check the recycling code at the bottom of your plastic before tossing it in the bin, and consult your local officials about specific recycling requirements in your community.

re-purpose your plastic

Did you know that plastic can help your garden grow? Plastic cups and egg cartons make great nurseries for seedlings, and large plastic beverage containers with the  bottoms cut off make perfect plant protectors. For sanitary reasons, it's not a good idea to refill water bottles for drinking, but plastic bottles can be used to make birdfeeders, slug catchers, and even piggy banks!

But why stop there? Recyclable odds and ends, such as plastic lids and Styrofoam packing peanuts, can become clever holiday ornaments or children's crafts, and glass jars—once washed—can hold everything from pennies to peanut brittle.

With a little thought and creativity, there's no end to how you can extend the life of your recyclables long before they hit the bin.

get mileage from your shoes

Leather shoes might seem like the natural choice, but there's nothing natural about the chemicals used during the tanning process. Before you buy your next pair of shoes, consider footwear made with certified organic or all-natural materials such as hemp or organic cotton.

Extend the life of the leather shoes you already own by conditioning the leather often and getting them re-soled as needed.

Did you know that old, worn-out athletic shoes with synthetic or rubber soles can be recycled? Check with your local retailers about shoe recycling programs, or visit *www.runtheplanet.com* and *www.nrc-recycle. org/reuseashoe.aspx* to learn about numerous shoe recycling options.

donate your clothes

Breathe new life into old clothes by giving them away! Why let clothes you no longer wear or need languish in the back of your closet? Remember: if the clothing doesn't fit, donate it!

You will find dozens of local charities and non-profit organizations that would welcome your tax-deductible gift. Your own family members, friends,  and neighbors might be thrilled to receive your "vintage" hand-me-downs, especially items like winter coats, shoes, and children's clothing. And who knows? Maybe they'll return the favor.

If you have new, formal, or designer pieces, you might consider "donating" them to a consignment store.

simplify your life

Searching for the simple life? Get rid of your stuff—the green way.

You can streamline your life *and* reduce waste in our landfills at the same time!

Purge your home of all the items you no longer need or want, and recycle, sell, or donate them to a worthy cause. Post items online at *eBay* or *craigslist,* or advertise in your local newspaper. Have a yard sale.

Contact charitable organizations, schools, friends, and neighbors and ask them if they could use whatever you've got.

 Or simply sign up for *Freecycle* and join the millions of people who are "changing the world one gift at a time."

recycle your cell phone

Keep your old cellular phone and other electronics out of landfills by recycling them with your town or designated recycler.

Cell phones contain mercury and lead that could present a hazard to you and your immediate environment over time.

Most cell phone retailers, as well as private companies, will buy or recycle your old cell phones. You can also donate your old phones to various nonprofit organizations.

Electronics equipment should also be recycled or donated. Check with *http://earth 911.org* to find a recycling center or donation location near you.

swap your books and magazines

Avid readers rejoice! By swapping your books and magazines with friends and neighbors, you can double up on your reading while helping to reduce waste. What a great way to "subscribe" to multiple magazines and read new bestsellers and old classics! You can also participate in various online book swapping groups such as:

www.paperbackswap.com
www.bookmooch.com
www.bookins.com

Of course, your local library, hospital, or senior center would welcome a donation of reading materials, so do consider passing along your print treasures when you're finished with them.

pack your garbage can

Reduce your trash by crushing it down to its true size.

By maximizing the space both inside your garbage bags and your trash can, you will use fewer plastic bags and streamline your garbage. Why use two garbage bags when your compacted trash will fit inside one?

You will also improve the efficiency of the trash collection process and reduce carbon emissions at the same time. The less time a garbage truck spends idling curbside during trash pick-up, the less fossil fuels consumed.

use biodegradable bags

Whenever possible, choose earth-friendly "plastic" bags for your trash.

Biodegradable, starch-based garbage bags derived from corn, such as BioBag *(www.biobagusa.com)*, are excellent alternatives to the traditional, petroleum-based plastic bags that dominate our landfills. When exposed to the earth's elements and micro-organisms in an "open" landfill, biodegradable bags will begin to decompose like other natural materials and leave no harmful residues behind.

pick up litter

It may not be your litter, but it's our planet. And you can help lead the way to a greener earth through your example.

Pack a bag or two, and make a habit of picking up trash while you walk the dog or stroll through the park, and dispose of it the right way. Recycle plastics, cans, and old newspapers. Cash in returnables and keep the change!

Share the fun and multiply your efforts by enlisting the support of family and friends. Invite them to participate in town beautification projects, or declare your own neighborhood clean-up day.

dispose of toxins the right way

Harsh toxic chemicals do not belong in landfills, and you can do your part by disposing of them the right way.

The first step to managing your chemical waste is being able to identify it. Do not throw things like batteries, paint, wood stain, motor oil, and oven cleaner in the trash. Here's a good rule of thumb: *When in doubt, don't throw it out.*

Instead, call your local waste management division and ask for guidelines regarding the disposal of your household chemicals.

In the future, seek out nontoxic product  alternatives that require no special handling or disposal. (*See* pages 72-74, 78, 82, 84, and 86.)

compost your organics

Whether you live in a city apartment or house in the suburbs, composting your organic waste has never been easier.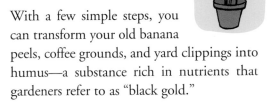

With a few simple steps, you can transform your old banana peels, coffee grounds, and yard clippings into humus—a substance rich in nutrients that gardeners refer to as "black gold."

Humus adds moisture and conditioning to soil, and encourages healthy growth of both indoor and outdoor plants.

Use a pre-made compost bin, a black plastic garbage bag, or a homemade receptacle, and start composting today! Find *The Complete Guide to Composting* online at *www.compost guide.com*.

fix leaks

A leaky faucet or running toilet can waste thousands of gallons of water over the course of a year, and that's no drop in the bucket!

Save water and reduce unnecessary waste by making sure faucet handles are turned completely to the off position. If your sink or shower head drips, or your toilet hisses and hums while continuously refilling its tank, you're wasting more than precious water. (Have you seen your water bill lately?)

Make today the day you dust off that tool box and find a wrench, or call a plumber to fix those leaks.

install water saving devices

Simple devices installed on your faucets, toilets, and shower heads can significantly reduce your water and energy consumption.

Shower heads that gush like waterfalls are luxuries our earth cannot afford, but modern faucet aerators, low flow toilets, and  energy saving shower heads designed to limit the flow of water coming out of your tap will preserve our most precious resource without sacrificing water pressure.

Your five-minute shower using a water saving device can be just as effective and satisfying while consuming 20 percent less water.

And less hot water circling your drain translates into smaller energy and water bills.

drink tap water

Buy a filter and drink from the tap.

The bottled water industry drains natural resources and pollutes our earth. According to The Green Guide *(www.thegreenguide. com)*, Americans consume more than 70 million bottles of water each day—last year adding 22 billion empty plastic bottles to our landfills. And a year's production of the plastic water bottles uses up 1.5 million barrels of oil!

When you consider the huge waste of natural resources and the quantities of greenhouse gases emitted in the production, manufacturing, and distribution of bottled water, you might go running for the tap.

use the dishwasher

Put the sponge down and step away from the sink.

An energy-efficient dishwasher (look for the Energy Star label) requires less than five gallons of water to clean a full load of dishes, so why are you still washing by hand?

You will save time and thousands of gallons of hot water each year by loading your dishwasher to capacity and allowing it to do its job. And you already know the equation: less hot water equals lower energy and water bills.

Of course, whether it's the dishwasher or the washing machine, you should only wash full loads to save water and energy.

scrape your plates

Do not pre-rinse your dirty dishes before loading them into the dishwasher.

You will save hundreds of gallons of water each week by simply scraping your plates. Deposit any food scraps into the garbage or compost, and load your dirty dishes as they are. Your dishwasher will take it from there and use less water in the process.

For items you must wash by hand, such as fine china or heavily soiled pots and pans, plug the sink and reuse rinse water where possible to allow dirty dishes to soak.

take shorter showers

Reducing the amount of time you spend singing in the shower can save you money while conserving water.

Shower with a purpose: *to get clean.*

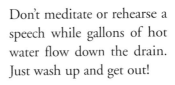

Don't meditate or rehearse a speech while gallons of hot water flow down the drain. Just wash up and get out!

If you have a water-saving shower head equipped with an on/off switch, you can flip the switch to "off" after getting wet and lather up at your leisure. Better yet, why not shower with a friend?

By all means, keep singing in the shower. Just sing shorter songs—or make it a duet!

turn off the tap

One of the most obvious ways to conserve water is to simply turn off the tap.

Every day we perform basic rituals that involve running water. A minor adjustment to the way we approach these rituals can have a major impact on our overall water consumption.

 Think about how much water you could save if you turned off the hose while washing your car, or shut off the faucet while shaving, washing your hands, rinsing vegetables, and brushing your teeth.

Water conservation. It's every bit as easy as twisting a knob.

use a rain barrel

For the uninitiated, a rain barrel is a receptacle for collecting rainwater that attaches directly to the downspout of a gutter on a house. Rainwater can be used to water indoor plants and outdoor lawns and gardens, and even to wash your car.

And wherever you live, rainwater collects fast! One inch of rain falling on an area of 1,000 square feet will yield about 600 gallons of rainwater.

With a specially designed rain barrel or just a large bucket, you will have plenty of water for your plants and other household needs without ever reaching for the hose.

reuse dirty water

Think twice before you pour "dirty" water down the drain.

Remember: one person's waste water is another's waterfall!

The water you use to rinse dirt off your farm-fresh vegetables may be perfectly suited to your flower garden. The same goes for stale water left inside sports bottles or canteens, water from the baby's bath, or water collected inside a dehumidifier.

Depending on the concentration of residual minerals or starches, you should dilute water that has been used to cook pasta or rinse out the coffeepot before watering your plants.

water at dawn and dusk

Water your plants at dawn and dusk for maximum efficiency.

Americans use about one-third of all residential water to nourish their lawns and gardens (depending on their location within the  country), but unfortunately, much of that water gets wasted due to over-watering, evaporation, and runoff. Poorly designed landscaping features present additional challenges to water-conscious home gardeners.

Set your sprinkler system for the times of day your vegetation can best absorb water, or personally see to the watering yourself first thing in the morning and at the end of the day.

Of course, you should not water your lawn or garden during a drought.

kill weeds kindly

The sight of unwelcome weeds in flower and vegetable gardens can provoke violent reactions in even the most peaceful of gardeners. We yank weeds up by the roots and whack them into submission, but occasionally, we seek out something stronger.

Fortunately, you can kill weeds kindly using a green thumb and nontoxic remedies.

 Pour scalding hot water directly on your weeds, or choose from dozens of all-natural weed management products that do the job without any harsh chemicals. Ordinary white vinegar mixed with a pinch of salt and liquid dish soap, for example, zaps weeds without contaminating the environment, and it's cheaper!

nourish plants naturally

Choose eco-friendly fertilizers for your lawn and garden.

Commercial fertilizers made with toxic chemicals derived from fossil fuels do more than enhance the appearance of your plants. They can seep  chemical toxic waste containing lead, cadmium, mercury, arsenic, and dioxin into the soil and groundwater, endangering the entire ecosystem. And nitrogen runoff wreaks havoc in the environment.

All-natural fertilizers, such as humus from a compost bin or grass clippings from your lawn, will promote healthy growth without the use of dangerous chemicals. Consult your local garden supply store for simple solutions and other green alternatives.

plant hardy varieties of grass

Plant types of grass that are best suited to where you live.

Don't attempt to recreate a golf course green on your front lawn without first checking with your local nursery or garden club to learn about which grass varieties thrive in your soil and climate. Of course, you should resist the temptation to plant high-maintenance, fickle, or disease-prone varieties.

In mountainous or desert regions, consider alternatives to grass. An artful array of native shrubs, trees, flowers, and ground cover can look just as nice and requires no mowing.

use a push mower

Manage your lawn the green way: use a push mower!

Most suburban homeowners do not require fossil fuel-consuming ride-on tractors or automatic mowers to maintain their lawns.

Technological advances have made modern push mowers the natural choice for the average homeowner. They're cost-effective, easy to use, and good for the environment. Better yet, they help keep you looking as neat and trim as your lawn!

With a little human power and a push mower, you can have the greenest lawn in town. What are you waiting for?

sharpen your mower blades

Keep your grass green and healthy by trimming it the right way.

Sharpen your mower blades to cut grass without tearing it unevenly and risking damage to the plant. A clean cut keeps grass healthy and green from the roots to the tips, and healthy grass requires less water.

Taller grass provides more shade to roots and also helps to maintain moisture better than short grass, so set mower blades as high as you can manage.

And for the greenest of lawns, leave the clippings to decompose and return vital nutrients to the soil.

xeriscape your lawn

Xeriscaping features well-designed landscapes that conserve water while celebrating the beauty of native, drought-resistant species.

Instead of planting grass and exotic vegetation in your yard or garden, plan your landscape design around plants and trees that are already adapted to your soil, temperature, and climate, and group them strategically.

Plants with similar water needs should be planted together, as should plants that serve as repellents for certain insects. French marigolds, for example, repel insects that can harm tomato plants, and both garlic and onions, independently, are effective against many insect garden pests.

plant trees strategically

Did you know that trees can lower your carbon output *and* your electric bill?

Plant deciduous trees (trees that lose their leaves in the winter) on the south and east sides of your home, and evergreen trees to the north and west.

During the hot summer months, leafy green trees can block the sun and provide valuable shade to your house and air-conditioning units.

In the wintertime when trees are bare, the warming rays of the sun will help heat your home, while evergreen trees will provide a natural barrier against cold northerly winds.

add mulch and compost

Reduce your water use by adding compost and mulch around your plants.

Compost, either from your compost bin or from the garden supply store, adds vital nutrients to the soil and encourages healthy plant growth. Healthy plants require less water.

The addition of mulch, such as cedar or pine, further insulates plants and helps the soil retain moisture, so your plants will require even less water. As an added bonus, mulch lends a neat appearance to your plant beds and discourages the growth of weeds.

cut the lights

Turn off the lights when you leave a room.

Lighting represents the biggest energy drain in the average American household and business, so try to get into the habit of turning off the lights whenever you exit a room.

During daylight hours, make the most of natural light by placing mirrors and other reflective objects near windows to catch the light and diffuse it throughout your home.

After dark, be mindful of the lighting you do use. Consider installing a dimmer switch to regulate lighting. Instead of leaving bulbs burning all night long, try a solar-powered night-light.

set a timer

Don't waste energy by leaving the garage light on all night long.

Illuminate driveways and walkways only when necessary by setting a timer on your outdoor lights.

You can also install lights with motion sensors that save energy by providing light when movement is detected. As an added benefit, motion sensor lights provide safety, too, by alerting you to movement outside your house.

Better yet, see the light with solar lighting. Harness the sun's energy by day, and use solar pathway lights, spotlights, and motion detector lights after dark without relying on any electric power. Brilliant!

change your lightbulbs

 This is one of the easiest and most effective ways to reduce your energy consumption. Switching to compact fluorescent lightbulbs (CFLs) will save you money—to the tune of about $90 per bulb, over the life of the bulb, for each $5 CFL you use.

Energy-efficient CFLs last up to 10 times longer than standard incandescent bulbs and require 75 percent less electricity. If every American household replaced just one incandescent bulb with an Energy Star CFL, it would save enough energy in one year to light three million homes, and prevent greenhouse gases equivalent to the emissions of more than 800,000 cars.

Note: CFLs contain trace amounts of mercury, but they're still a better choice than standard lightbulbs, which emit more mercury during their production. Check with your CFL retailer and your town about disposal of CFLs.

dust!

Here's a bright idea: Use a feather duster to cut your energy consumption!

A simple flick of the wrist can improve the efficiency of your lighting. By dusting off your lightbulbs, you will improve their energy efficiency and double their brightness. And with all the extra light emanating from your clean bulbs, you should be able to shut off a light or two.

But why stop there? You can clean your way to a greener home one appliance at a time. Dust or vacuum computer equipment, heating vents, radiators, and refrigerator coils to maximize their effectiveness and keep them in top working order.

decorate with LED lights

Choose energy efficient LED lights (light emitting diodes) for your holiday decorations.

The popular incandescent mini lights we use to decorate our Christmas trees and homes each year consume huge amounts of energy and generate enough heat to pose a fire hazard. In comparison, LEDs require a fraction of the energy and stay cool to the touch. A mini LED light consumes about .04 watt of electricity whereas a typical mini light uses .45 watt.

LEDs are approximately twice the cost of incandescent lights but last much longer—the equivalent of 10 years of continuous indoor use.

unplug your appliances

Did you know that your electronic devices still draw energy when they're plugged in but not in use? This is called standby power.

This means that the appliances and electronics you leave plugged in, such as your recharger, coffee-maker, microwave, TV, audio equipment, and computer, could be drawing up to 40 percent of  the energy they require when you're actually using them.

By simply getting into the habit of unplugging these devices when they're not needed, you will reduce your energy consumption and unnecessary carbon emissions.

replace old appliances

Zap your electric bill by updating your major household appliances with new, energy-efficient models.

Look for products bearing the Energy Star label to reduce your electricity consumption and save money. A front-loading washing machine, for example, requires 60 percent as much water and energy as older, top-loading models. Energy-efficient dishwashers, refrigerators, and even computer screens all add up to make a difference in your overall carbon footprint.

clean your filters

The dust, lint, and debris that accumulate on filters inside your household appliances will hinder their performance and make them work twice as hard.

Save energy by cleaning them from the inside out. Remove lint from your dryer filter after each load of laundry. Empty the pool filter regularly. Replace the filters on air-conditioning units and heating systems as needed. Remove lint from the back of electronics equipment, and even your hair dryer.

Think about it: with a clean blow dryer, you'll not only decrease the amount of time you spend drying your hair, but you'll also cut your energy use. As a side benefit, your air will be cleaner as well.

shut the fridge

Don't cool the kitchen by keeping the refrigerator door open while you contemplate what to eat.

Each time you open the door to your fridge or freezer, cold air escapes and your refrigerator cranks into high gear, using extra  energy to cool things back down. To maintain a consistent temperature of 36-38°F in your fridge and 0-5°F in your freezer, minimize menu planning while the doors are open.

The reverse is true for your oven. Remember: a watched pot never boils. Avoid opening and shutting the oven to prevent heat loss and allow your food to cook.

wash in cold water

Keep your bright colors bright and your carbon output low by washing your clothes in cold water.

Hot water not only requires energy to heat, but it can be tough on certain fabrics and shrink clothing. Your laundry will get just as clean when washed at cold or warm settings, and your clothes will be brighter and more durable for a longer period of time.

The energy you save using cold water translates into dollars and sense: that is, dollars in your pocket and good sense for our earth.

air-dry your stuff

Rethink the way you dry your laundry and dishes.

 Instead of using the clothes dryer, line dry your laundry. Hang wet washables outside on a clothesline if weather permits, or use a space-saving indoor drying rack. Your clothes will dry faster than you think without any help from the electric company.

Washing dishes in your dishwasher is more energy-efficient than washing by hand, but not so with drying. Turn off your dishwasher when it reaches the drying cycle, and either hand dry your dishes or pull out the racks to let them dry naturally.

charge your cell phone in your car

Get charged up—*by your car*.

Charge your cell phone while on the go using energy from your car battery. You will enjoy the same fully charged cell phone, but from an energy source that regenerates itself from the car engine.

 Cell phone chargers drain energy from electrical outlets while they charge your phone. And if you leave your charger plugged in after you've finished using it, it continues to draw unnecessary energy.

Eliminate all this energy waste
and get charged up the right way
by relying on a regenerating,
"renewable" source.

lose the leaf blower

Gather your leaves the old-fashioned way: with a rake!

 The once peaceful annual ritual of raking leaves has been reduced to little more than gas-powered noise pollution, thanks to the ever present leaf blower. And, sadly, the din doesn't stop there.

Homeowners and landscaping crews alike use fossil fuel-powered blowers from dawn until dusk all season, blowing leaves and debris into manageable piles. Once the first snowflake hits the driveway, out come the snow blowers for more of the same.

Restore peace and quiet to your neighborhood by losing the leaf blower and using human power to clean up your yard.

get an energy audit

Evaluate your home's energy efficiency and identify corrective measures by conducting a home energy audit.

You can hire a professional to assess the energy efficiency of your home, or you can ask your local power company to do it—probably at no charge.

A home energy audit examines all the ways your home consumes and wastes energy, including the efficiency of heating and cooling systems, the settings on your thermostat and hot water heater, and the effectiveness of weatherproofing windows, outlets, and vents. A thorough review of your home provides the opportunity to make small changes that can add up to big energy savings.

dress for the weather

Don't use your household thermostat to regulate your body temperature. Wear layers!

You will save energy and money by reaching into your closet before you reach for the thermostat.

And remember: the layered look never goes out of style. Just add or remove a layer of clothing to maintain a comfortable body temperature all year long. When the weather heats up, peel off a layer. When you feel a chill in the air, grab a sweater!

install a programmable thermostat

Customize your home heating and cooling system with a programmable thermostat.

Programmable thermostats are easy to install and allow you to significantly curb your energy consumption.

Set your thermostat to a comfortable temperature—68°F or lower in the winter and 78°F or higher in the summer—and schedule it to go on only when you need it. For example, you may want the heat to go on in the early morning and again in the evening, but not overnight while you sleep. In the summertime, you may desire air-conditioning at night but not during the day.

insulate your home

Cut your heating and cooling needs by properly insulating your home.

Start with the obvious: windows and doors. Install caulking and weather-stripping to fix leaks and drafts, and use window treatments like insulated curtains and shades to help maintain indoor temperatures.

Seal air leaks around heating vents and electrical sockets, and insulate your water heater, all accessible hot water pipes, and the few feet of your cold water inlet pipes.

regulate your water heater

Set your water heater at a reasonable temperature.

Many homeowners unwittingly set their water heaters too high, a waste of both energy and money. A temperature of 120°F is high enough to kill bacteria and address your household needs.

Of course, you should take proper care to insulate your water heater and pipes to avoid unnecessary heat loss. You can buy a fiberglass tank wrap kit at your hardware store and install it yourself easily.

reverse ceiling fans

Keep air moving in the right direction by reversing your indoor ceiling fans by the season.

In the summer, the fan should draw cooler air upward, reducing the need for air-conditioning. In winter, a ceiling fan should push warm air back downward.

Since different manufacturers use different fan designs, the only way to tell which way to aim your fan blades is to turn your fan on high, stand under it, and decide which way is more comfortable. Then keep it that way for that season and reverse it for the other season.

forget the fireplace

Don't let your indoor heat go up in smoke!

Not all fireplaces are created equal, and in fact, many standard fireplaces will send the heat inside your home right up the chimney.

In the end, your cozy night by the fireplace may actually increase your energy needs and home heating costs.

If you're looking to lower your overall carbon footprint, research the most efficient ways to heat your home, and certainly con-sider options such as wood burning stoves.

That said, use your standard fireplace more for ambiance than heat, and be sure you remember to close the flue when you're through.

don't use space heaters

Think twice before you plug in that space heater.

Despite their widespread use, space heaters do a poor job of generating and maintaining heat.

And though relatively inexpensive to buy, space heaters can be ultimately more expensive to own and operate due to their inefficiency and energy requirements.

Take measures to properly insulate your living space, and explore energy efficient heating options that will prove to be more cost-effective over time.

go solar

Consider harnessing solar power for your home and water heating needs.

 Solar power represents a viable alternative to fossil fuel-dependent home heating systems, relying on little other than the sun for energy. In sunny climates such as the Southwest, solar panels may indeed be the natural choice.

Some states offer incentives and rebates that make solar systems even more accessible to the average homeowner. Long-term financing from lenders is another option, providing a solar hot water or power system for a low monthly fee.

Depending on where you live, a solar power system might be the right choice for you.

consider geothermal heating and cooling

Geothermal heat, or heat from the earth, is a renewable source of clean energy produced without the aid of fossil fuels.

Geothermal heat is an excellent alternative to heating and cooling systems that depend on gas, oil, or coal.

Also known as a ground source heat pump, geothermal heat is generated by the circulation of liquid in closed loops of pipes buried underground. A heat exchanger transforms this liquid to heat, which can then be used to warm the air inside your house. Conversely, a geothermal system can also be used for cooling in the summer.

consider wind power

Wind power is a clean, renewable energy source of growing popularity around the globe.

Traditional windmills harnessed wind power to pump water or process grains. Nowadays, large-scale wind farms built on flat open areas convert wind energy to electricity, reducing greenhouse gas emissions associated with electricity derived from fossil fuels.

Consumers and small businesses can use smaller, residential wind turbines to complement solar power systems or to offset utility costs.

Consider installing a residential wind turbine as a means of reducing your carbon output and your electric bills.

avoid the oven

When possible, use the microwave instead of the oven to save energy.

Electric and gas ovens produce excessive heat and require large amounts of energy to cook foods. By comparison, a microwave requires far less time and energy to heat food sufficiently.

When you do use the oven, plan accordingly. Bake several dishes at the same time, and to prevent the heat from escaping avoid opening and closing the oven door. When food is nearly cooked, turn off the oven and allow the residual heat to finish the job.

downsize your pan

Conserve energy in the kitchen by choosing the smallest pan possible to suit your cooking needs.

Smaller pans heat up more quickly and require less energy than larger pans, especially when you match the size of the pan to the size of the burner. The pan should cover the burner and extend no more than one inch beyond the surface.

With a small pan, you will keep your cooking time to a minimum and, when you add a lid to your pot or pan, you will reduce your overall energy use by about 20 percent.

consider a solar oven

If you live in a hot, sunny climate, consider using a solar oven to prepare cooked foods the "green" way.

Solar ovens are portable, easy-to-use ovens that harness the sun's rays to cook food. When placed outside in a sunny location, they heat up within a matter of minutes and cook food over a period of hours—anywhere from two to four hours or longer depending on the size of the meal.

Foods cooked in solar ovens require no additional water and retain more nutrients than foods cooked in traditional ovens.

use cloth towels

Opt for cloth over paper whenever possible for your daily household needs.

The production of disposable, bleached paper products kills trees and releases harmful carcinogens called dioxins into the environment.

Why contribute more toxins to our world and fill landfills with paper goods when we can reuse cloth products for the same purposes?

For household cleaning tasks, extend the life of your old clothes by transforming them into cleaning rags. Instead of paper napkins and towels, invest in a colorful assortment of reusable cloth hand towels.

say no to aerosols and solvents

Avoid the use of aerosols and solvents and choose nontoxic, green alternatives whenever possible.

 Household solvents may include products like nail polish remover, paint, and even perfume, and containers should be considered hazardous waste and disposed of accordingly.

Spray-on products sold in aerosol cans (hairspray, deodorant) contain chemical propellants such as hydrocarbons, carbon dioxide, and nitrogen that contribute to greenhouse gases and smog.

Buy home and personal care products in pump sprays instead of aerosol cans, and opt for all-natural, nontoxic alternatives.

clean green

Choose earth-friendly, nontoxic cleaning products for your home, or make them yourself.

Here are some tried-and-true "green" cleaners:

- Glass cleaner
  *1/4 cup white vinegar or
  1 tablespoon lemon juice,
  2 cups water*

- Scouring powder
  *baking soda and salt, or
  baking soda with lemon juice
  or vinegar*

- Disinfectant spray
  *1/2 cup eucalyptus or
  peppermint oil
  1 gallon water*

clean green

- Furniture polish
  1/2 cup white vinegar
  1 teaspoon olive oil

- Laundry softener
  1/4 cup baking soda

- Drain cleaner
  equal parts baking soda and salt,
  flushed with boiling water

- All-purpose cleaner
  1/8 cup borax
  1 quart hot water

- Grease remover
  1/2 teaspoon washing soda
  2 tablespoons white vinegar
  1/4 teaspoon liquid soap
  2 cups hot water

service systems regularly

Schedule regular service checks on home heating systems to maintain the system's efficiency and to protect the air inside your home.

Professional maintenance of home heating equipment such as furnaces and gas heaters will reduce the likelihood of malfunctions  that can release dangerous fumes into the air you breathe. If you have a fireplace and use it with some frequency, have it professionally cleaned by a chimney sweep at least every other year.

Of course, you should install smoke and carbon monoxide detectors near your furnace and bedrooms, and replace the batteries annually to ensure their proper functioning.

pot a plant

Use houseplant to freshen the air you breathe at home.

 Indoor air pollution caused by chemicals in building materials, new carpets, paint, and household cleaners, to name just a few, might have more of an effect on our health than previously thought.

According to NASA and the Associated Landscape Contractors of America, the following indoor plants are especially effective at purifying the air: bamboo palm, Chinese evergreen, chrysanthemum, corn cane, dracaena, English ivy, gerbera, peace lily, snake plant, and philodendron.

Find more indoor plant recommendations at *www.zone10.com/tech/NASA/Fyh.htm.*

keep a handle on humidity

Prevent the growth of potentially dangerous mold and mildew in your home by managing ventilation and humidity.

> Excessive humidity can spawn naturally occurring molds and mildews that, when left unchecked, can have harmful health consequences.

As a precaution, use a dehumidifier to dry up any humid areas inside your home, such as your basement, and keep the air circulating with well-maintained ventilation systems.

It's also a good idea to have your indoor air quality tested by a professional to determine the presence of any unwanted substances.

avoid commercial air fresheners

Freshen the air inside your home without the use of chemical-laden sprays and deodorizers.

Instead of plugging in a commercial air freshener or dousing your furniture with "fabric refresher," fill a pot with water, dried cloves, cinnamon sticks, and orange rind and let it simmer on the stovetop.

 Essential oils, such as orange, lemon, or peppermint, can diffuse pleasant, unobtrusive aromas into the air. Add a drop to furniture polish composed of olive or vegetable oil, and give your tabletops a quick swipe. Or dab the surface of a lightbulb and allow the aroma to diffuse throughout the air.

light green candles

Fill the air with all natural aromas by burning candles made of pure beeswax or soy wax.

Burning petroleum-based candles with synthetic fragrances and bleached wicks containing lead may do more than create ambiance in your home. You may unwittingly release dangerous toxins and chemical fumes into the air that can be hazardous to you and the environment.

Beeswax and soy wax candles scented with sweet honey and essential oils burn twice as long as paraffin wax candles without the risk of polluting your indoor air. Remember to trim your candle wicks to 1/4 inch to reduce excess soot.

check your cookware

Know what's really cooking by taking a closer look at your pots and pans.

At very high temperatures, the coating on many popular brands of nonstick cookware can degrade, exposing you and the environment to harmful chemicals. Other dangerous chemicals are released into the environment during the manufacture of nonstick coatings such as Teflon.

Avoid using nonstick cookware at prolonged high temperatures, and contact your local waste management division about recycling pans with obvious chips or peeling surfaces.

Choose cast iron, stainless steel, glass, or ceramics, and look for earth-friendly nonstick cookware that's safe for you and for the environment.

hire a green contractor

Think green before you embark on your next construction or home improvement project.

Find a contractor who has been certified by the U.S. Green Building Council's Leadership in Energy and Environmental Design
(LEED) program. A "green" contractor will help maximize your home's efficiency by identifying areas of improvement and making changes that will reduce energy waste and consumption. Structural enhancements to floors, walls, ceilings, doors, and windows, combined with upgrades to new, energy-efficient appliances and systems, can translate into huge savings over time. And the use of eco-friendly materials and products can help improve your indoor air quality.

use low or no-VOC products

Choose low or zero-VOC (volatile organic compounds) products for your household needs. Many paints, finishes, adhesives, and even carpets emit formaldehyde and other chemical pollutants into the air inside your home.

Improve the quality of the air you breathe by using only low or zero-VOC water-based paints and products without added chemicals.

If low or zero-VOC products are not an option, choose water-based latex paints over solvents, or ask your paint retailer about other eco-friendly options.

explore alternatives to wood

Consider alternative building materials when embarking on a new construction project.

Research products made with recycled materials, such as building composites made of reclaimed wood and plastic or insulation composed of recycled newspapers or plastic.

Always research the origin of the materials you are buying, and seek out local manufacturers whenever possible.

If you choose wood, support sustainable forestry by buying wood that has been certified by a third party, such as the Forest Stewardship Council.

embrace essential oils

Repel insects without the use of potentially dangerous chemicals.

Insect repellents that contain DEET and other chemicals may harm more than mosquitoes. Before you apply an insect repellent containing DEET to your skin or release it into the natural environment, consider the alternatives.

 Products with an active ingredient derived from all-natural essential oils, such as geranium or lemon eucalyptus oil, have proven to be safe and effective remedies against biting insect pests.

Certain plants, such as the citronella bush, serve as natural insect repellents, and can be planted near outdoor patios as a deterrent.

set traps

Use simple, nontoxic traps to manage insects and other pests.

Instead of setting out poisonous edibles or spraying chemical toxins on surfaces in and around your home, set up simple traps to catch pests.

First, secure all cracks and gaps in your cabinets and floors to deny entry to marching ants and other crawling pests.

Sticky flypaper is very effective at ensnaring flying insect pests. For mice and rats, consider a catch-and-release rodent trap if you live near an open area. In the garden, entice slugs into shallow glass jars filled with an inch of beer.

try natural remedies

Combat invasive pests with natural remedies.

Ants will retreat from several common substances, including fresh lemon juice, black pepper, boric acid, talcum powder, and ordinary dish soap. Locate the point of entry and use any one of these ordinary substances to stop ants in their tracks.

For ticks and fleas, use an herbal "tea" repellent made with one cup fresh or dried rosemary and about two quarts of boiling water. Wash pets with ordinary soap and water, dry them thoroughly, and then apply cooled rosemary-water with a spray bottle or sponge and allow it to dry. Reapply as often as needed.

the eco-friendly office

When one tugs at a single thing
in nature, he finds it attached
to the rest of the world.

—JOHN MUIR

Going green at the workplace makes good business sense.

Companies that adopt sustainable business practices can *reduce, reuse, and recycle* their way to higher profits and reduced costs—both for themselves and our earth.

A green strategic plan will incorporate ways in which employees can minimize their consumption of natural resources and reduce unnecessary waste. Simple changes to the daily routine can have a significant impact over time.

Imagine the paper *you personally* could save by printing documents on both sides. Just think how many Styrofoam coffee cups you could keep out of landfills each year by switching to a reusable mug! How much energy could you conserve by turning off your lights and shutting down your computer each day?

Now multiply these actions by everyone in your office.

Your collective efforts can make all the difference to the future of your business and our environment. Work smart. Take one green step today.

power off at the end of the day

Treat your office as you do your home, and turn off the lights and electronic equipment when you leave.

Establish a policy whereby the last person to leave the office at night is responsible for turning off the lights. This simple act will significantly reduce unnecessary energy waste. But don't stop there.

Promote energy conservation and awareness by turning off and unplugging all non-

essential electronic equipment at the end of each workday. Shut down computers, printers, and copy machines. Unplug the coffee maker and the microwave.

switch to energy star computers

Investing in new technology throughout the office can save you big!

Most new laptop computers with the Energy Star designation, for example, are about twice as energy efficient as models produced just one year ago. And in general, laptop computer systems consume less electricity—not to mention space—than older, standard computer workstations.

An up-front investment in new, energy-efficient technology will pay for itself in no time in the form of lower utility costs.

work from home

Explore telecommuting options at your workplace and do the math.

If you drive to work, you can save nearly 40 percent of your transportation costs by working from home just two days per week.

If everyone took advantage of telecommuting options, just think of the results: vastly reduced energy consumption, less pollution, less traffic, and so on.

Encourage your employer to consider the benefits of telecommuting, or consider alternative modes of transportation, including public transportation and carpooling, if your work does not lend itself to telecommuting.

opt for e-mail

Reduce unnecessary paper waste and conduct your business the green way.

Modern technology allows us to share vast quantities of information with the touch of a button, eliminating the need for excessive copying and printing, so use it to your advantage.

E-mail business communications to colleagues and clients, and file them electronically. Reach customers with targeted e-blasts instead of expensive direct mail or paper-driven campaigns.

Conduct business meetings online; make presentations across the globe using little more than a laptop computer; and share documents and meeting minutes without printing a single copy on paper.

file documents electronically

Find a new use for those large metal filing cabinets, because thanks to new technology, they're no longer needed.

Unless your paper files are mandated by law, create and store documents online in folders on a shared computer network server where they are accessible by your business colleagues. Use password-protected files to store personal and confidential information on your computer.

Electronic data storage will improve the efficiency of your office by making it easier to retrieve and store information. And it will reduce the costs and waste associated with paper, printing, and copying.

think before you print

Preview your documents before you send them to the printer, and print only as many pages and copies as you need.

Too often, we bypass the "print preview" option on our computer screens and commit documents to paper before they're fit to print. In haste, we subsequently waste paper, ink, time, and energy trying to print documents correctly.

Improve your energy efficiency and use of resources by taking advantage of online editing tools and proof-reading your work. And whenever possible, print and copy on both sides of the page to avoid unnecessary waste.

fill the blue bins

Recycle in the office just as you do at home.

An active workplace recycling program demonstrates a commitment to environmentally friendly practices and promotes a reduce/recycle/reuse way of thinking in all employees.

Make a personal commitment to recycle at every opportunity. Deposit recyclable plastics, glass, paper, and aluminum into designated recycling receptacles, and encourage your colleagues to do the same.

 Most of the 67 tons of paper that Americans use each year ends up in landfills. Take advantage of the sea of office waste paper by reusing it as scrap. You'll never need sticky notes again!

reuse office supplies

Treat office supplies with care and respect, and reuse them as many times as possible.

Circulate important mail or internal documents to work colleagues in reusable envelopes. Find new uses for boxes, packing materials, and padded envelopes. Repurpose old file folders, and cut outdated letterhead and paper supplies into scrap notepaper.

Establish an office repository for used writing utensils, staplers, tape dispensers, and so on, and encourage staff to look there first before taking new products or materials.

recycle ink cartridges

Always recycle used computer, printer, and copier ink cartridges.

These days, companies that produce ink cartridges for home and office use make recycling very easy. Old, empty ink cartridges can be repackaged in special boxes or envelopes they supply, many times with the return address already printed in place and the postage paid.

What could be easier? Make a habit of repackaging empty ink cartridges to the manufacturer's specifications at the same time you replace a new cartridge.

streamline supplies

Keep your office supply stock to a minimum to avoid waste.

Let's face it. The greater the supply of pens, sticky notes, paper clips, and notepads, the greater the human tendency to hoard. Hoarding results in unnecessary waste of both resources and energy—particularly when you consider the fossil fuels required in the manufacturing, packaging, and distribution of supplies.

Break the cycle of over-consumption and curb your appetite for extra highlighter pens and manila folders. Your restraint is good for business and the environment, so take only what you need.

buy recycled paper products

Demonstrate a commitment to the environment by purchasing 100% recycled paper products for your home and office. If your employer does not already do so, make the suggestion and encourage your colleagues to do the same.

 You can find a wide variety of recycled, chlorine-free paper products of excellent quality such as copy paper, letterhead, sticky notes, and file folders. Always look for products labeled 100% recycled, and check for the certification label of the Forest Stewardship Council (FSC) indicating that the paper originated in forests managed in an environmentally responsible manner.

choose nontoxic pens and adhesives

Avoid office supplies and products made with chemical solvents.

Instead of adhesive tape and solvent-based glues like rubber cement and hobby glue, choose nontoxic alternatives to suit your purposes. Staples, paper clips, nontoxic glue sticks, and white glue work just as well and do not release unwanted chemicals into the environment.

Likewise, avoid colored markers with chemical solvents and opt for nontoxic, water-based varieties. Or just use crayons or colored pencils.

use a plain paper fax machine

Choose energy-efficient office equipment that produces minimal waste.

If your work processes involve daily use of a fax machine, make sure you use a plain paper machine with paper and ink cartridges that can be recycled.

 Some model fax machines use special fax paper that cannot be recycled. This means that all unwanted fax documents you receive on this special paper, as well as faxed documents you no longer need, end up being tossed in the trash and taken to the dump.

do dishes

Keep a supply of reusable dishes and serving ware in the office kitchen or break room, and use them!

Restore civility to the workplace during lunch or break time. Drink tea from an actual teacup! Eat your lunch from a ceramic plate or bowl! Enjoy the luxury of silverware! Dab your lips with a cloth napkin!

And while you're washing dishes after lunch, you can reflect on how many disposable paper and Styrofoam products you and your green colleagues kept out of the landfill that week.

brew coffee with a reusable filter

Green up the office coffeepot!

Start by lobbying for *real* coffee—that is, 100 percent organic, fair trade, shade-grown coffee.

When you brew quality coffee using a reusable mesh or cloth filter instead of paper, you give new meaning to "the daily grind."

Raise the bar even further by serving it up in reusable mugs with all the usual accompani-

ments, and fewer of your colleagues will feel compelled to detour to the local coffeehouse on their way to work—a habit that is both expensive and incredibly wasteful.

pack a mug

If you're like many Americans whose daily routine involves buying coffee "with wheels" (aka "to go"), then make sure you bring your own mug.

Many popular coffeehouses offer a discounted price to customers with their own cups, another excellent incentive to avoid consuming unnecessary resources and producing waste in the form of paper or Styrofoam cups. The same offer usually extends to drinkers of tea and fountain beverages.

Make a habit of bringing your favorite mug or travel cup with you every day. It will help you stay hydrated and refreshed while reducing paper waste.

bring your lunch

Instead of buying takeout meals every day for lunch, bring your lunch from home in a reusable container.

 Diet and budget reasons aside, you should avoid takeout meals because of the packaging. Styrofoam cups and containers are not often recyclable and take centuries to decompose in landfills. Your daily takeout meal may give you a severe case of indigestion when you consider your personal contribution to the town dump.

Make a conscious effort to reduce your disposable waste. If you must buy takeout, take it out in your own container.

encourage eco-conscious catering

Reduce unnecessary waste when ordering catered foods for work-related meetings and events.

Don't over-order extra trays of food that will languish in the conference room or kitchen and ultimately be thrown in the trash. Be realistic and order only what you need. When the meeting has concluded, spread the wealth. Promptly notify all staff of the availability of catered leftovers, and encourage them to help themselves.

Dishes and glassware may not be logistically possible for catered meals for larger groups, so seek out disposable paper products made with 100 percent recycled materials.

out and about

*Take care of the earth
and she will take care of you.*

—AUTHOR UNKNOWN

The choices you make every day have a great impact on the future of our planet. And as a consumer, the choice is yours.

From the vehicle you drive to the food and products in your shopping cart—not to mention your vacation lifestyle and choices—you become either part of the problem or part of the solution.

Making green choices in the marketplace can be confusing, as not all that glimmers green is earth-friendly, certified organic, and recyclable.

Indeed, it's not always easy being green.

But a green consumer is an educated one. Arm yourself with information before you buy. Read labels. Ask questions. Pay attention to things like packaging and ingredients you can't pronounce. Consider the long-term environmental effects of the "convenient" disposables we toss in landfills *by the billions*: diapers, napkins, razors, water bottles. The choice is yours. Make it a green one.

avoid short trips

Before you hit the road, make a plan.

Short car trips waste fuel and are best avoided, especially when your car engine is cold. A cold engine may use nearly twice as much fuel and not become efficient until about five miles into your trip.

Accomplish as many quick errands as possible by other means—such as on foot or by bike—and save the others for a longer excursion.

You will save time and gas by strategically planning your route according to the errands you need to accomplish.

drive the speed limit

Put the pedal to the metal, and pay the price.

This may sound obvious, but driving at higher speeds consumes more fuel—as does quick acceleration. Furthermore, driving above the speed limit also places you at greater risk of accidents.

The more fuel you require and use, the greater the price paid by you and our earth.

Manage the fuel in your gas tank more efficiently by driving at slower speeds and with a light foot, not a lead foot. You will save both energy and money, and possibly your life.

avoid idling

An idling engine wastes fuel and pollutes the environment.

Contrary to popular belief, it is more fuel-efficient to turn off your car engine when you're stuck in a traffic jam than to let your engine idle. Of course, you will also reduce carbon emissions by limiting the time your engine runs.

On cold mornings, idling to warm up the car burns nearly twice as much fuel and may cause unnecessary  wear on your engine. Modern engines require only about 30 seconds of idling when started in cold weather.

inflate your tires

Did you know that tire pressure can affect fuel consumption?

When you operate your vehicle on tires that are not inflated to the manufacturer's recommendations, you risk more than just accidents and malfunction. You pay more because your engine works harder, consumes more gas and, in so doing, expels more carbon dioxide into the environment.

Avoid unnecessary fuel consumption and potentially dangerous driving conditions by checking your tire pressure regularly.

get long-lasting treads

Keep your tires on the road and out of landfills by investing in long-lasting treads.

Despite concerted recycling and retreading efforts as well as state regulations prohibiting the dumping of used tires, about one quarter of all tires end up in landfills or are illegally dumped, polluting our environment and posing fire hazards.

Before you buy new tires, check out the possibility of retreading the tires you own. If that's not a possibility, buy tires with extra durability and monitor their wear. Of course, if you have scrap tires in your possession, contact your local waste management division about recycling options.

remove bike racks and accessories

Improve your car's fuel efficiency by streamlining its exterior.

Bike racks and other rooftop or tailgate accessories interfere with your car's aerodynamics, making your engine work harder than necessary, and consume extra fuel to get you to your destination.

When these accessories are not being used, make the extra effort to remove them. The energy you expend taking off the ski rack will be rewarded with improved fuel efficiency.

take the stairs

Instead of forcing your way into a crowded elevator every day, take the stairs!

Do your part to cut back on energy use by avoiding the elevator whenever possible. Walking up 60 flights of stairs may be unrealistic, but if you live or work closer to ground level, break the habit of using fossil fuel-derived power to get you where you're going. Bypass the elevator button and step it up—and down—the stairs. You'll improve your health and reduce energy consumption in the process.

hit the streets

Put a spring in your step by relying on your own two feet to get you to your destination. You'll save money, reduce carbon emissions, and get a workout to boot!

If you're fortunate enough to live within walking distance of your workplace, then, by all means, walk to work. What better way to start and end your work day than by taking in the sights and breathing some fresh air?

A daily stroll can be incredibly invigorating, especially if you walk with a purpose. Why not pack a bag and pick up litter along the way?

ride a bike

Got wheels? Ride 'em!

When you've got local errands to run or places to go, consider hopping on your bicycle before you burn up the road and environment with fossil fuels from your car.

Rediscover the joy of riding a bike,
or pick the wheels of your choice.
Strap on some rollerblades and
feel the wind on your face!
Hop on a scooter!

However you roll, do so with the environment in mind.

drive a hybrid

Drive smart by considering a fuel-efficient vehicle that uses clean energy technology.

Hybrid cars combine a gas-fueled combustion engine with a battery-powered electric motor, resulting in improved fuel efficiency and decreased carbon emissions. Some hybrids have an electric-only drive capacity at low speeds and while idling, while others switch from one power source to the other at stoplights and during stop-and-go driving.

Look for a wider selection of plug-in hybrids and other alternative fueled vehicles on the horizon, including cars that run on biodiesel, a blend of diesel fuel and vegetable oil that generates fewer and less toxic emissions than regular diesel.

share a ride

Let someone else do the driving!

Whether you're traveling to work or to a local concert, carpool! Share a ride with colleagues or friends to reduce the number of cars clogging our roads and spewing dangerous carbons into the environment.

> Did you know that for every gallon of gas saved, carbon dioxide emissions are reduced by 20 pounds?

Better yet, take public transportation. Hop on a bus. Ride a subway or train. Ride your bike. Whatever you do, avoid making solo trips in your car routinely.

shop with a purpose

Make a shopping list before you hit the mall.

Buying products you *want* but don't *need* fuels a dangerous cycle of production and consumption. The more we buy, the more we produce. The more we produce, the more we buy. And the production, distribution, and transportation of goods and services translate into the burning of fossil fuels and increased carbon emissions.

Break the cycle of over-consumption. A simple shopping list will not only help you save money, but it will help save the planet.

be an educated consumer

Educate yourself so you can make environmentally conscious buying decisions.

Make an effort to understand what you are buying. Take the time to read. Seek out information from every available source: people, online resources, books, newspapers, and magazines.

Pay attention to food and product safety recalls, expiration dates, and warnings and disclaimers on the products you buy.

You're in control, and a little knowledge will help you see through false and misleading advertising claims and lead you to make wise—and green—purchasing decisions.

buy in bulk

Avoid excessive packaging by buying in bulk.

Americans produce on average about 4.5 pounds of trash a day, much of it composed of the plastic, Styrofoam, and cardboard packaging from the items we buy. This trash winds up in landfills and takes years to decompose.

 Do your part by choosing products designed in eco-friendly, low-impact packaging, and recycle relevant plastics, cardboard, and glass. Whenever possible, buy in bulk, and steer clear of products with packaging that creates unnecessary waste, especially items packaged for single use such as individual snacks, drinks, and pre-packaged lunches.

buy local

Buy food and goods produced by local vendors.

Did you know that a thriving local economy actually helps the environment?

By supporting local organic farmers and companies by buying their goods, you reduce the need for other products to be shipped across the globe. This cuts back on all the energy and materials required to manufacture, distribute, and transport these goods to you.

Buying locally produced goods not only encourages a sustainable way of life, but it saves natural resources and reduces carbon emissions as well.

buy organic, earth-friendly products

Fill your shopping cart with greens!

Buy organic foods and natural fibers (e.g., organic cotton, hemp, and wool) produced without the use of pesticides harmful to you and our earth. Chemical pesticides used in conventional farming leach into waterways and disrupt the natural ecosystem, and traces of these same chemicals can be found on the nonorganic produce you buy at the store.

Choose products with minimal packaging, and avoid Styrofoam and nonrecyclable plastics whenever possible. Check the recycling code number on the bottom of a product to identify whether or not you can recycle it. (*See* pages 133-134.)

read labels

Don't fall prey to clever marketers and false advertising claims.

Read labels on the products you buy. Find out where an item was made, and take time to decipher ingredient lists. Some polysyllabic ingredients may be as difficult to swallow as they are to pronounce.

Thanks to tricky food labeling laws and unregulated industries, designations like "green" or "all natural" don't necessarily mean what you think.

Look for certifications or certain key words such as "certified organic" or "100% recycled" that indicate a product was produced without harmful pesticides from 100% recycled materials.

buy secondhand

Next time you're in the market for something new, consider secondhand stuff.

Remember: If it's *new to you*, it's still new. And by buying used merchandise, you are helping to reduce the environmental costs associated with the manufacturing and distribution of new products.

 Check out your local consignment, thrift, and secondhand retailers to find all kinds of treasures: furniture, housewares, clothing, artwork, jewelry.

Make the rounds at community tag sales, or find used items for sale online at places like *eBay* and *craigslist*. Better yet, register online at *Freecycle* and get items for free!

use reusable bags

Switch to reusable cloth shopping bags.

When offered the option of paper
or plastic bags at the checkout
counter, avoid both by bringing
your own reusable bag made of
organic cotton canvas, hemp, or
some other durable, washable fabric.

Plastic bags are petroleum based and can take
500 years or more to decompose. Paper bags
not only kill trees, but they require fossil fuels
to produce and distribute. Keep both out of
our landfills by packing your own bags.

remember your bags

Bring your reusable shopping bags
with you to the grocery store.

This might seem fairly obvious, but the
reality is that many well-intentioned people
routinely forget them. Often, it's not until we
hear the words, "Paper or plastic?" that we
remember the location of our "green" bags.

Help yourself develop new habits when you
shop. Buy reusable nylon bags that fold up
into handy little squares, and carry them with
you inside a purse or pocket. In addition,
store your organic cotton canvas or hemp
shopping bags in a visible location in your
house or vehicle.

opt for cloth

Save trees and the environment by opting for cloth over paper whenever possible.

Switch to cloth napkins and hand towels instead of buying disposable paper goods. While you're at it, forgo the disposable diapers and choose cloth.

On average, one person consumes about 2,200 paper napkins each year. Give up your napkin habit and save trees, eliminate waste, and conserve energy!

By choosing cloth, you can prevent billions of tons of paper waste from entering our landfills, while saving forests that harbor wildlife and purify the air we breathe.

buy bleach free recycled paper products

Look for paper products labeled "100% recycled" and "totally chlorine free" (TCF) or "processed chlorine free" (PCF).

The processing of many commercial paper products involves the use of chlorine bleach, causing carcinogens known as dioxins to be released into the environment.

Do your part by buying eco-friendly paper products made without dangerous chemicals, and always buy recycled.

According to Seventh Generation, if every U.S. household replaced just one 500-count package of napkins with 100% recycled ones, we could save 2.4 million trees, 6.3 million cubic feet of landfill space, and 887 million gallons of water.

buy plastics by the number

Look for the recycling arrow symbol and number at the bottom of the plastics you buy, and find out what types of plastics are recyclable in your area (usually numbers 1 & 2).

Here is a list of plastics by the number.

- **1** Polyethylene Terephthalate:
  Soda and water bottles,
  cooking oil bottles

- **2** High Density Polyethylene:
  Detergent bottles, milk
  and juice jugs

- **3** Polyvinyl Chloride (PVC):
  Plastic pipes, shrink wrap, some
  food and detergent containers

- **4** Low Density Polyethylene:
  Dry cleaning bags, produce bags

buy plastics by the number

- **5** Polypropylene:
  Drinking straws

- **6** Polystyrene:
  Styrofoam cups,
  packaging peanuts,
  to-go containers

- **7** Other (includes
  Polycarbonate, Acrylic,
  and Fiberglass):
  Food containers, baby bottles

avoid number 3 & 7 plastics

For the sake of your health and our planet's well-being, do not use products identified as Number 3 or 7 plastics.

Number 3 plastic, polyvinyl chloride (PVC), emits chemical pollutants into the environment throughout its life cycle and has been linked to serious health effects, including cancer. PVC is used in plumbing pipe, some water bottles, outdoor furniture, and other items.

Number 7 plastic, a catchall category that includes polycarbonate and acrylic, is used in baby bottles, water bottles, and food storage containers. Number 7 plastic has been linked to hormone disruptions and reproductive problems.

avoid disposable products

Adopt a sustainable lifestyle by avoiding single use products.

Many of our products of convenience, such as Styrofoam cups, disposable razors, and "throw away" cameras, are not so convenient for our environment. We toss single use products in the trash as fast as manufacturers can produce them, filling our landfills and polluting the environment with toxic waste.

Break the cycle of production and waste by investing in reusable, rechargeable, quality products.

They will cost less over time, last longer, and reduce carbon emissions from all the disposables you're not using and discarding. How convenient!

shop for natural fiber

Conventional cotton, the world's most popular fabric, has a dirty little secret: it's hooked on pesticides. In fact, the cultivation of cotton accounts for 25 percent of the world's pesticide use. Add to that the harsh chemical bleaches and dyes used in processing the fabric, and cotton may be even worse for the environment than petroleum-based synthetics.

Thankfully, we have several eco-friendly options to the fabric dilemma, including: **wool**; **organic cotton**, a soft, allergy-free fabric without pesticides, bleaches, or harsh chemicals; **hemp**, a strong, versatile all-natural fabric; and **bamboo**, one of the fastest growing plants in nature and popularity.

buy recycled jewelry

Indulge yourself with recycled jewelry and help the planet at the same time.

The mining of precious metals and gemstones wreaks havoc on the environment, adding dangerous toxins to the earth and destroying ecosystems in the process.

 You can help by buying previously owned gold, diamonds, and other jewels from jewelry and vintage stores as well as the online marketplace. You can also use your own broken jewelry or pieces you no longer want to create beautiful new designs. Consult your local jeweler for ideas or seek out companies such as *www.greenkarat.com*, offering jewelry that is ecologically and socially responsible.

beware chemical contaminants

Pay attention to the ingredients of the cosmetics and personal care items you use every day, and direct your consumer power accordingly by buying healthful products.

In the United States, many common hair and skin care products such as shampoos, hair dyes, and moisturizers contain chemicals that have been banned elsewhere in the world due to their risk to people and the environment.

Phthalates, industrial chemicals found in many products including nail polish and deodorant, have been linked to birth defects and should be avoided, as should coal tar, diethanolamine (DEA), and parabens, among others.

beware chemical contaminants

Before you endanger yourself and the environment, research your products at *www.cosmeticsdatabase.com* or *www.safe cosmetics.org,* and use only those with 100 percent natural and safe ingredients.

support fair trade

Buy products that have been labeled "fair trade."

Fair trade companies pay their suppliers a fair wage for products and services, and do not employ the use of child labor.

When you buy products bearing the "fair trade" label, such as coffee, tea, cocoa, bananas, and cotton, you are assured that the people who produced your goods—from farmers to factory workers—received fair remuneration for their services.

join a farmers' cooperative

Support your local organic farmers—and *eat fresh, delicious foods*—by joining a farmers' cooperative—also known as Community Supported Agriculture (CSA).

Local organic produce is far superior to supermarket fare; fruits and vegetables from distant lands are harvested early in order to make the trip from the farm to your table.

Organic farmers do not use harmful pesticides on their crops, and you can taste the difference. As part of a farmers' cooperative, you usually receive weekly rations according to what is in season. Take your produce home in reusable cloth or mesh bags, and you're in for a treat.

know the dirty dozen

Learn which fruits and vegetables are most susceptible to pesticide contamination, and always buy them organic.

Conventionally grown produce may look succulent and fresh, but it can contain residues from more than 50 different pesticides. Some fruits and vegetables absorb these toxins more readily than others, posing a potential risk to you and your environment.

Support safe farming practices and buy organic produce whenever possible, especially the following twelve pesticide-prone fruits and vegetables: **strawberries, raspberries, imported grapes, peaches, nectarines, cherries, pears, apples, spinach, potatoes, bell peppers**, and **celery**.

drink green spirits

Make a toast to the health of our planet with organic wine or beer!

 Organic wine is made from 100 percent organically grown grapes. Unlike commercial wines, organic wines do not contain additives that can cause allergic reactions in some people, nor do they have any traces of pesticides, herbicides, and insecticides that could harm you and the environment.

Organic beer is another pure spirit made with certified organic ingredients. Organic hops and grains grown without chemical pesticides and fertilizers are brewed to perfection, creating tasty, earth-friendly brews.

eat your greens

A vegetarian diet may save more than your health—and that's not just a load of hot air.

By all accounts, the news really *stinks*. Methane, one of the "natural byproducts" of the livestock industry (e.g., the gaseous emissions from cows, pigs, and chickens during the digestion process), is one of the largest sources of greenhouse gases on the planet.

When you consider the amount of water, energy, and natural resources required to raise livestock, and the cost to you and our environment, plant-based protein sources start to look mighty good. Veggie burger, anyone?

be responsible

Do right by your family and pets by acting responsibly on their behalf.

As the adult, you're in charge, and it's up to you to provide a safe, nurturing environment that promotes your family's growth and well-being for years to come.

 The lifestyle choices you make today—from the foods you prepare and the products you buy to the ways in which you protect our earth's natural resources—will shape the lives of your loved ones and the earth they inherit.

A "green" lifestyle is a responsible one. Be responsible, and make eco-friendly choices that promote life.

teach your children well

Model the type of behavior you want your children to emulate.

Demonstrate your commitment to being a responsible global citizen by teaching the next generation from an early age how to live a green life.

Show them how to recycle and explain why it is important.

Educate them about natural resources, and illustrate ways to conserve energy, water, and trees.

Give them hands-on experiences. Walk them to school! Visit state parks, demonstration farms, and local landfills! Camp! Plant a garden!

Teach your children well and they will develop a deep, abiding respect for their planet.

wash your children safely

Keep your children clean and safe by using all-natural products.

Babies and children absorb chemicals through the skin far more readily than do adults, so it's especially  important to check the ingredients of the skin and hair care products you use on them.

In the United States, many popular products designed for babies and children contain perfumes, dyes, and other additives that may cause allergic reactions.

Keep these chemicals away from your children and out of our ecosystem by choosing fragrance-free products made with all-natural or organic ingredients.

choose cloth diapers

Use washable cloth diapers instead of disposables.

Cloth diapers have come a *long* way, with new, easy-to-use, cute, and colorful styles made from the softest organic cotton. They're durable, cost-effective, and by far the greener choice when you consider the alternative.

Despite claims that cloth diapers use more water and energy to launder, studies have shown that they have a much lower impact on the environment than do disposables.

Disposable diapers represent the third largest source of solid waste in our landfills and take up to 500 years to decompose. And each year, we toss *18 billion* more in the trash.

buy eco-friendly toys

Spoil your children with natural, eco-friendly toys.

Choose simple, timeless toys made from nontoxic, organic materials like wood and fabric that are stimulating to growing brains and safe enough to nibble.

 Look for eco-friendly toys made by local artisans: pull toys, puzzles, dolls, games. Buy nontoxic art supplies and recycled paper. Research toys by brand and category online to ensure their safety.

Avoid synthetic, mass-produced products that may contain hazardous substances such as lead or phthalates (PVC). Look for a "PVC-free" label on plastic toys, and beware products manufactured overseas that may not adhere to U.S. standards.

give green gifts

Give the earth a gift by giving green gifts.

Choose eco-friendly products made from all natural, organic materials, and reduce the ecological impact of your gift-giving by opting for consumable gifts, such as flowers or a basket  of assorted local, organic treats like honey, jam, maple syrup, wine, and chocolate.

Give gifts that keep on giving: a share in a farmers' cooperative, a donation in someone's name to an important cause, a carbon offset, plants and trees.

Of course, creatively wrap your gifts with low-impact, recyclable, or reusable materials such as newspaper or cloth bags.

buy rechargeable batteries

Invest in a renewable source of energy to power your gadgets: rechargeable batteries.

 You'll pay a little more up front to replace your standard batteries with rechargeables, but realize significant savings over time. So will our earth.

Each year, nearly three million batteries are dumped in landfills, leaking mercury into waste matter in a gigantic toxic stew.

Keep mercury out of our landfills by disposing of old batteries and other toxic waste the right way (go to *www.earth911.org* to search for a disposal site near you), and get charged from a renewable energy source.

use a chalkboard

Make a commitment to reducing your paper waste and protecting our natural resources by seeking alternatives to paper whenever possible.

Hang a community message board in your home to record messages and leave notes. You will keep your family just as organized and informed by writing information on a chalkboard or a white board instead of paper notepads or sticky notes.

If you can't break the paper habit, then write your notes on the backsides of used envelopes and waste paper before you recycle them.

buyer beware

Seek out certified organic, all-natural products and materials for your home.

Unbeknownst to many consumers, certain chemicals found in common household products and furnishings may present a risk to you and your environment.

 Flame retardant chemicals applied to mattresses and carpets have been shown to leach into our indoor air and bloodstreams. And products manufactured in foreign countries may not adhere to U.S. guidelines concerning lead and other toxic substances.

Research the products you buy, including mini-blinds, ceramics, paint, and toys, and test the ones you own with a kit or by hiring a professional.

support green business

Seek out green services in the market-place and at home.

Choose dry cleaners that employ nontoxic, green cleaning methods. Patronize restaurants that specialize in local, organic foods. Hire green lawn service providers and household helpers. Support shops that sell locally-produced goods made with all-natural or recycled products.

As the consumer, the choice is yours. Make it a green one.

plan ahead

Get unplugged before you take off on your next trip!

If you're planning to be away from home for days or weeks, make sure you reduce your energy use while you're gone. Safety reasons aside, you will save energy and expense by unplugging appliances large and small, such as your toaster, coffeemaker, washing machine, and dryer. Unplug your electronics while you're at it, including the television, DVD player, computer, and printer.

Of course, be sure to adjust your home heating and cooling systems to their most energy efficient settings, and don't neglect your water heater.

take an eco trip

Do you want to make a difference for our planet? Are you interested in a little adventure? Take an eco trip!

Volunteer your time and assist on a wide variety of environmental projects and opportunities across the globe.

Pick a spot on the map and get your hands dirty by helping out on an organic farm. Explore the wilderness as a volunteer with the  National Park Service and other environmental organizations.

Or take an eco trip closer to home. Contact local environmental groups, educational institutions, and private green enterprises and see how you can support their efforts.

find a green hotel

Check in at a green hotel next time you're away from home.

Eco-conscious hotels take measures to cut greenhouse gas emissions and conserve natural resources. A green hotel utilizes nontoxic cleaning supplies, recycled paper products, and energy-efficient lighting and appliances.

 As a guest at a green hotel, you can do your part to save water and energy. Keep laundry-related costs down by using the same linens throughout your stay. Refrain from cranking up the heat or air-conditioning or pocketing extra toiletries. And turn out the lights and TV—just as you do at home.

recycle on the go

Bring your good habits with you wherever you travel.

Vacations don't excuse you from the green lifestyle you embrace at home. Be a responsible global citizen by showing the same regard for our earth's natural resources as you do in your own backyard.

Pay attention to the water and energy you use. Avoid buying products with excessive packaging or those packaged for individual use. 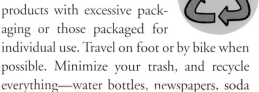 Travel on foot or by bike when possible. Minimize your trash, and recycle everything—water bottles, newspapers, soda cans—in the proper location.

tread lightly

Challenge yourself to lower the environmental impact of your next vacation by choosing an alternative mode of transportation.

 Use your own two feet to get you where you're going and walk! Savor the sights on a leisurely stroll through city streets or quiet parks. Pick up the pace and jog along distant coastlines and residential neighborhoods. Hop on a bicycle and explore new territory.

Avoid renting a car if at all possible, and instead opt for public transportation. If you must travel by car, make sure you rent a fuel-efficient model, or perhaps a hybrid to create less pollution.

speaking up, getting down

Never doubt that a small group of thoughtful, committed citizens can change the world; indeed, is it the only thing that ever has.

—MARGARET MEAD

L ive green, learn the scene, and talk it up! Magnify the impact of your individual efforts to help the environment by spreading the word.

Share your successes with others. Start an office or community bulletin board with tips on how to *reduce, reuse, and recycle.* Educate your friends about earth-friendly choices and how to make simple changes in their daily routines.

Become informed. There is a world of resources out there, a mouse click away.

Learn, model, and communicate.

Become a political force by joining environmental and activist organizations, signing

petitions, attending events, writing and calling your politicians, and working to bring about change in your community and beyond.

lobby your leaders

Challenge your local leaders to make the environment a priority.

Find out where they stand on issues that affect your town and our earth. Ask what they are doing to cut carbon emissions, conserve natural resources, and educate the public in your community.

Learn the facts. How will the town reduce its reliance on fossil fuels? Will open space be protected? What happens to hazardous waste? Are there any green programs in place?

Hold public officials accountable and don't be afraid to ask the tough questions.

vote smart

Support political candidates who make the environment a priority.

Our planet needs help. Greenhouse gas emissions due to the burning of fossil fuels threaten devastating climate change. Water resources are being depleted faster than they're being replenished. Chemical toxins simmer in overcrowded landfills the world over.

Each member of our global society plays a role in helping to reverse the damage. Strong political leaders can effect positive change on a large scale by enacting eco-friendly legislation and tough new regulations—and holding people and businesses accountable.

buy carbon offsets

Invest in clean energy by offsetting your personal carbon emissions one metric ton at a time.

When you buy a "carbon offset," you give money to an organization that will pursue clean energy projects to help negate carbon emissions.

For example, an organization such as *www.carbonfund.org* or *www.terrapass.com* may develop alternative energy sources like wind farms, or coordinate projects involving the large scale planting of trees.

Businesses and individuals alike can support these efforts and our planet by buying carbon offsets as well as pursuing their own conservation efforts.

support green institutions

Support organizations that demonstrate a commitment to the environment.

Schools and businesses that promote environmental stewardship incorporate eco-friendly practices in their daily operations. Look for established recycling and composting programs, and LEED-certified building techniques and designs. Routine business activities will be environmentally sound and utilize clean energy technologies.

 Support these institutions, and learn from them. Educate others about their cutting-edge environmental concepts and programs. Most of all, adopt environmentally sound practices in your daily life.

do one green thing

Take a stand on the environment.

Whether it's challenging your family to recycle more and consume less, or taking measures to reduce the amount of junk mail you receive at home, *act*. Make the call. Write the letter. Participate in the discussion.

Do one green thing, and do it today. Recycle one aluminum can. Replace one lightbulb.

Remember: one thing leads to another. And if you do something rather than nothing, it could mean *everything* to the future of our planet.

If everyone did just *one thing* each day to help reduce the use of fossil fuels and conserve our natural resources, we could change the world.

resources

Global Stewards
www.globalstewards.org/ecotips.htm
Environmental tips and sustainable solutions
for a healthy planet.

The Intergovernmental Panel on Climate
Change (IPC)
www.ipcc.ch
Official website of the Nobel Peace Prize-
winning international organization working
to educate the world about man-made
climate change and promote efforts to
combat that change.

The Pew Center on Global Climate Change
www.pewclimate.org
Leading forum for objective research and
analysis, and for the development of policies
and solutions to address global climate
change.

Global Change at the National Academy
of Science
www.dels.nas.edu/globalchange
Overview of the global change activities of
the National Academies, from the advisers
to the nation on Science, Engineering, and
Medicine.

The United States Environmental
Protection Agency (EPA)
www.epa.gov
Official site of the EPA, with resources,
links, and comprehensive environmental
information.

The Union of Concerned Scientists
www.ucsusa.org
Website of the science-based nonprofit group working for a healthy environment and a safer world.

The Green Guide
www.thegreenguide.com
National Geographic Society's comprehensive guide to all things green.

Greenpeace
www.greenpeace.org
Website of the leading environmental activist group with comprehensive information on global environmental challenges and ways to get involved.

Tree Hugger
www.treehugger.com
Comprehensive site promoting sustainability and all things green.

Earth911
http://earth911.org
Excellent environmental resource with links
to local recycling information.

Energy Star
www.energystar.gov
Joint program of the U.S. Environmental
Protection Agency and the Department of
Energy, focused on energy conservation.

just for kids

EPA Climate Change Kids Site
www.epa.gov/climatechange/kids
Environmental information, games, and
links for children and teachers.

The Pew Center on Global Climate
Change Kids Page
*www.pewclimate.org/global-warming-basics
/kidspage.cfm*
Climate change questions and answers,
links, and suggested actions to stop global
warming.

Time for Kids Special Report
on Global Warming
www.timeforkids.com/TFK/kids
Stories, games, and links about global
warming for children.